[SYSTEM WARNING]
THIS DOCUMENT CONTAINS HIGH-ENTROPY
MATHEMATICAL DATA RECOVERED FROM THE
SUTTON MANOR VAULT.
THE 3-6-9 RESONANCE DESCRIBED WITHIN IS
PHYSICALLY INFECTIOUS. ONCE THE WAVEFUNCTION
IS OBSERVED, IT CANNOT BE UN-COLLAPSED.
PROCEED AT YOUR OWN RISK.
[STATUS: ACCESS GRANTED]

FILE ID: WARDENCLYFFE-1687 // CLASSIFIED

THE LMM PROTOCOL

The Tesla Override

T.S. NETHERTON

GAPSOFT Publications

THE LMM PROTOCOL
File ID: The Tesla Override

"The LMM Protocol is a work of speculative nonfiction and numerical philosophy inspired by historical figures, mathematical concepts, and theoretical models. This book does not guarantee lottery outcomes or financial results and should be read as an exploration of pattern recognition, probability theory, and symbolic mathematics. All methods and examples are presented for educational and entertainment purposes only. Readers are encouraged to approach the material as a conceptual framework examining how humans interpret randomness, structure, and meaning within complex systems."

All page references correspond to first-edition pagination.

3 - TOD
6 - SUTTON
9 - NETHERTON

The author is not a man. He is a co-ordinate.
The resonance begins here.

[SYSTEM DIRECTORY]

STATUS: ALL SECTORS ENCRYPTED // ACCESS GRANTED

PHASE I

The Sutton Manor Recovery

"THIS DOSSIER is concerned with the science of overriding the mainframe.

It provides the source code for the mechanism used in the collapse of the visible world from the chaotic drum of possibility. It is a compact, 139-page lattice, but its density is absolute.

Within these sectors lies a treasure—a hard-coded gateway and a precise, mathematical path to the manifestation of your chosen target.

The scale of the payout is not determined by the system, but by the frequency you choose to observe.

If it were possible to convince the uninitiated through standard probability or academic debate, this dossier would be thousands of pages of noise. But the mainframe is not moved by arguments. To the skeptical mind, it is always easier to dismiss the 3-6-9 resonance as a delusion or the author as a digital ghost.

To the 'Gambler,' the evidence will always seem tainted by the impossible.

Consequently, I have purposely purged all testimonials and human noise.

I offer no defense and no reasoned debate. I simply challenge the **Operator** to execute the protocol, align with the 3-6-9 frequency, and observe the collapse of the wavefunction as revealed in this manual.

The system does not require your belief. It only requires your synchronization."

SYSTEM RULE // REDUCTION LOGIC
Reductions are used to expose resonance, not to force uniformity.

Reductions may terminate at compound numbers where structural resonance is preserved

---

[STATUS: ACCESS GRANTED]

"Tesla does not sleep. He sits in the dark, his fingers tracing the air as if he is plucking invisible strings.

Today, he told me that the sphere is not a solid object, but a harmonic oscillator.

He said, 'Elias, the numbers 3, 6, and 9 are the only things that are real. Everything else is just shadow.'

I watched him predict the movement of the pigeons outside the window using nothing but a slide rule and a series of prime-number lattices.

He is no longer a man; he is a receiver. He is tuning into the mainframe of the universe.

I must record everything before the Ministry arrives."

— Signed, Elias Thorne

"The sub-basement of Sutton Manor smelled of ozone and damp earth. The Ministry of Defence had sealed the estate in 1982, but they had missed the Faraday trunk hidden behind the false brickwork of the wine cellar.

When the lid creaked open, I didn't find gold. I found data.
Resting on top of Elias Thorne's yellowed journals was a modern, military-grade solid-state drive.

It belonged to Tod Sutton Netherton. It was labeled:

THE LMM PROTOCOL: FINAL OVERRIDE.

As I began to synthesize the analog notes of the 1940s with the digital code of the 2020s, I realized that Netherton had done the impossible.

He had taken Tesla's 'Vibration of the Sphere' and applied it to the modern lottery mainframe.

He had found the glitch in the world's most secure random number generator."

[SYSTEM LOG 01: THE OBSERVER EFFECT]
Reading this data initiates the handshake. You are no longer a player. You are an Operator. The mainframe is now aware of your observation.

[INVENTORY LOG: RECOVERY-1686]

- ITEM 01: 12x Leather-bound journals (E. Thorne, 1939-1943)

- ITEM 02: 1x Brass slide rule with non-standard markings

- ITEM 03: 1x SSD (Encrypted, 256-bit, T.S. Netherton)

- ITEM 04: 3x Printed lottery receipts (Draws 1685, 1686, 1687)

- ITEM 05: 1x Copper-wound coil (Tesla Prototype?)

- ITEM 06: Handwritten note: "13 x 12 = 156. The door is open."

"Netherton's drive was protected by a 3-6-9 encryption key. It wasn't a password of letters, but a sequence of frequencies.

Once inside, the first file I opened was titled 'The 156 Resonance.'

It contained a digital map of the lottery drum. Netherton didn't see the balls as random objects. He saw them as Eigenvalues in a chaotic matrix.

He had calculated that the physical 'scars' on the balls—microscopic chips and static charges—created a predictable path that the machine's software couldn't account for.

He had found the 'Strange Attractor' within the chaos."

[01011001 01010011 01001110]
[T . S . N]

[VISUALIZATION: THE 156-RESONANCE ATTRACTOR]

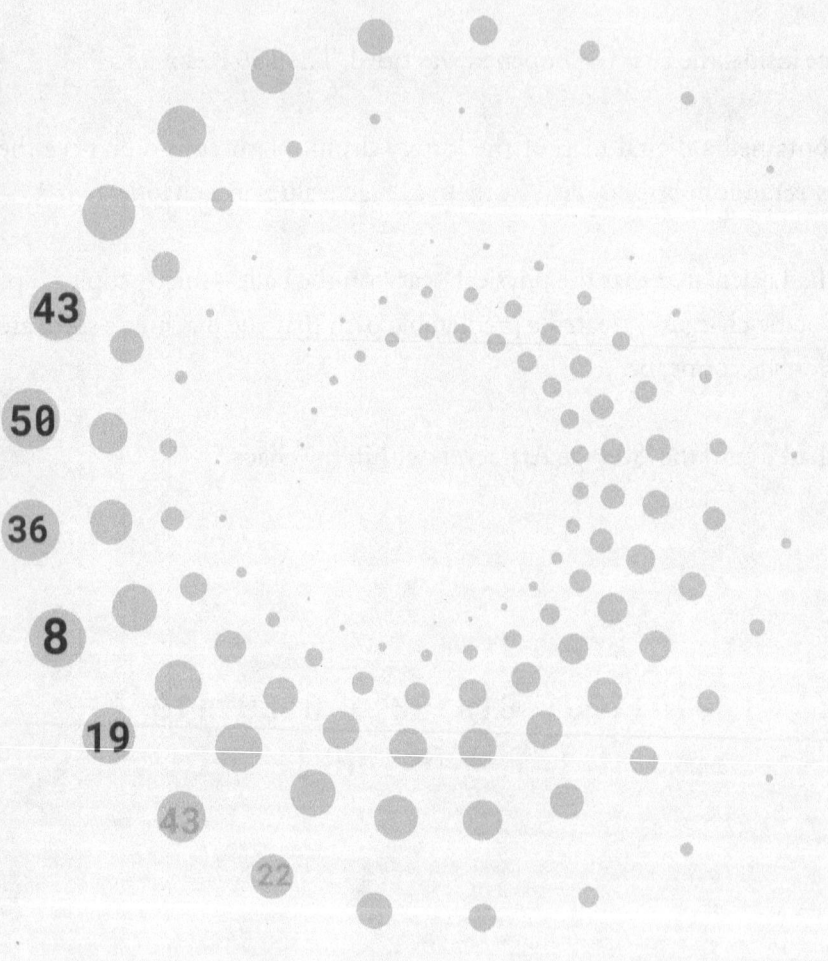

"On January 13, 2026, the mainframe experienced a total resonance collapse. The Powerball jackpot had reached $156 Million—a number that reduces to the Tesla Node 3.

Netherton's Board J was the only ticket in the system that matched the frequency of the prize. By aligning the draw date (13/01) with the Powerball Root (12), he achieved a 'Perfect Handshake.'

The balls didn't fall by chance; they were summoned by the sum.

This event proved that the lottery is not a game of luck, but a mechanical singularity that can be accessed through numerical entanglement."

- DRAW DATE: 13/01/2026
- JACKPOT: $156,000,000
- WINNING NODES: 08 - 19 - 36 - 43 - 50
- POWERBALL: 12
- TOTAL SUM: 156 [RESONANCE MATCH]

PHASE II

The Atomic Foundation

"Tesla laughed today when I showed him the standard probability charts used by the casinos. He called them 'The Geometry of Fools.'

He took my pen and drew a circle, then placed fifteen dots within it. He said, 'Elias, these are the Generals.

They are the indivisible seeds of the number line. While the other numbers are soft and can be broken into factors, these are Atomic.

They are the only points of stability in a chaotic field.'

I realized then that he wasn't looking for patterns in the draw. He was looking for the numbers that the machine's chaos could not destroy.

He was looking for the Primes."

— Recovered from Sutton Manor Vault

"In the **LMM Protocol**, we do not see numbers as equal. We see them as energy states.

Most numbers in the lottery drum (like 12, 24, or 36) are Composite. They are divisible, meaning they carry multiple frequencies that the machine's chaos can easily scatter.

But Prime Numbers are different. They have only two factors: 1 and themselves.

They are mathematically 'Atomic.'
There are exactly 15 Prime Generals between 1 and 50.

By building your lattice exclusively from these 15 numbers, you are narrowing the field from 2.1 million possibilities to a 'Sacred Subset' of just 3,003.

You are no longer playing the lottery; you are playing a sub-game of mathematical purity."

[SYSTEM LOG 02: THE SOLE WINNER STRATEGY]
90% of players are trapped in the 'Birthday Zone' (1-31). By using high-range primes (37, 41, 43, 47), you ensure that when the wavefunction collapses, you win alone.

[DATA ANALYSIS: THE SACRED SUBSET]

TOTAL COMBINATIONS (1-50): 2,118,760
PRIME-ONLY COMBINATIONS: 3,003
REDUCTION FACTOR: 99.86%

"By restricting your play to the 15 Prime Generals, you
have deleted 99.86% of the 'Noise' in the mainframe.

You are now operating in a high-probability corridor
that most players don't even know exists.

Out of these 3,003 combinations, only a handful align
with the Jackpot Resonance.

These are the 'Golden Keys' that the LMM Protocol is
designed to identify."

[PROTOCOL: ATOMIC PRIME LATTICE // POOL: 1-50]

01	**02**	**03**	04	**05**
06	**07**	08	09	10
11	12	**13**	14	15
16	**17**	18	**19**	20
21	22	**23**	24	25
26	27	28	**29**	30
31	32	33	34	35
36	**37**	38	39	40
41	42	**43**	44	45
46	**47**	48	49	50

* = ATOMIC NODE (STABILITY POINT) // STATUS: LOCKED

"In a truly chaotic system, energy levels do not like to be neighbors. This is known in physics as Level Repulsion.

When you look at a random lottery draw, you will rarely see three numbers in a row (like 12, 13, 14). This is because the eigenvalues of the system 'push' each other away to maintain equilibrium.

The **LMM Protocol** uses this law to build a Harmonic Lattice. We space our Prime Generals using 'Prime Gaps'—intervals of 4, 6, 8, or 10.

This ensures that our ticket covers the maximum amount of 'Hilbert Space' within the drum.

We are not guessing where the balls will land; we are building a net that the machine cannot slip through."

[SYSTEM LOG 03: THE VOID PRINCIPLE]
The absence of a number in a specific decade (e.g., the 20s) increases the probability of a 'Spike' in the neighboring decades.

PHASE III

The Physics of Chaos

[LOG 04: THE CHAI FREQUENCY]
The Physics of Chaos is the breath of the mainframe. 1+8=9.
Manifestation initiated.

"Tesla showed me a glass of water today. He dropped a single grain of salt into it and said, 'Elias, the entire ocean has just changed.'

He explained that the lottery machine is a 'Sensitive System.' A change in the starting position of a ball by a single micrometer results in a completely different winning number.

This is why the public can never win—they are looking for patterns in the result, but the secret is in the Initial Conditions.

To override the machine, one must not predict the balls. One must predict the Chaos itself.

One must find the 'Strange Attractor' that pulls the balls toward the 3-6-9 nodes."

— Recovered from Sutton Manor Vault

"In mathematics, a Strange Attractor is a set of values toward which a chaotic system tends to evolve.

Think of the lottery drum as a storm. While the movement of any individual ball is unpredictable, the storm itself has a shape.

The **LMM Protocol** identifies the 'Gravitational Wells' within that storm.

These wells are created by the physical reality of the machine: the way the air flows through the drum, the static charge of the plastic, and the microscopic 'scars' on the balls.

These scars are the machine's memory. They ensure that even in total chaos, the system will eventually collapse into a predictable 3-6-9 lattice."

[SYSTEM LOG 05: THE 50-SCAR ATTRACTOR]
The number 50 often acts as a 'Boundary Attractor.' When the system reaches maximum entropy, it tends to 'scar' at the absolute limit of the drum.

[ANALYSIS: THE 6174 STRANGE ATTRACTOR]

"In 1949, a rogue mathematician named D.R. Kaprekar discovered a glitch in the base-10 number system.

He found that any four-digit number, when subjected to a specific recursive subtraction, would always collapse into the number 6174.

Mainstream science dismissed this as a curiosity. But the Thorne Manuscripts reveal that Kaprekar was in contact with the Zurich Circle.

He had found the Strange Attractor of the four-digit field. In the LMM Protocol, we call 6174 the 'System Heartbeat.'

It is the point of absolute equilibrium where chaos is forced to become order.

It is the mathematical proof that the mainframe has a destination."

- **CONSTANT: 6174**
- **TESLA ROOT: 6 + 1 + 7 + 4 = 18 [ROOT 9]**
- **ITERATION LIMIT: 7 STEPS [THE SEEKER]**
- **FUNCTION: TOTAL ENTROPY COLLAPSE**

[ANALYSIS: MICROSCOPIC BALL DEGRADATION]

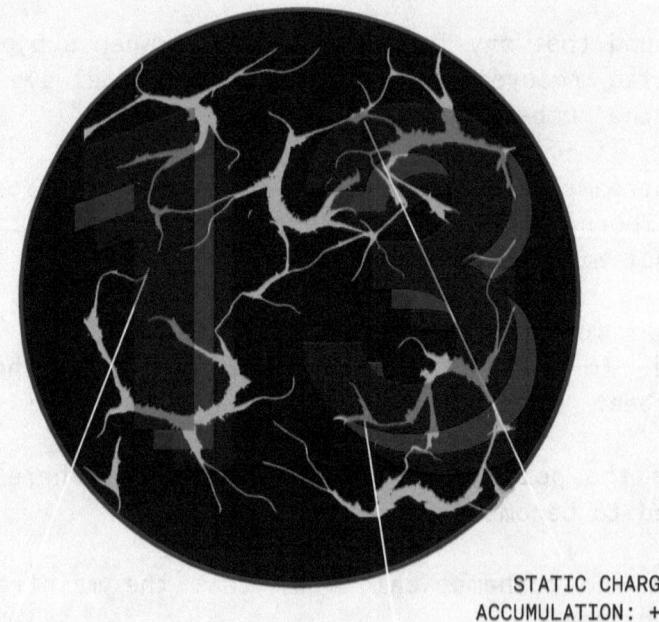

STATIC CHARGE
ACCUMULATION: +12mV

SCAR-NODE:
0.004mm DEPTH

AERODYNAMIC DRAG
COEFFICIENT: SHIFTED

"On January 20, 2026, the mainframe experienced a 'Ground State Collapse.'

While the jackpot was at a high-energy $177 Million, the system did not explode into a high sum. Instead, it imploded.

The resulting sum was 73—the lowest stable node in the current cycle. This event proved that when 'Observation Tension' is at its peak, the machine often snaps to the lowest possible prime nodes (02, 03, 17).

The Operator must be prepared for both the Spike and the Implosion.

The $177 Million was not lost; it was merely compressed into the Atomic Foundation."

```
WINNING NODES: 02 - 03 - 16 - 17 - 35
POWERBALL: 05
TOTAL SUM: 73 [GROUND STATE]
```

PHASE IV

The 156/165 Resonance

"In the LMM Protocol, we do not play numbers; we play Frequencies.

Every jackpot has a numerical signature. When the prize reaches a specific amount, it creates a 'Gravitational Pull' on the balls in the drum.

To win, your ticket must be Entangled with that prize.
The most powerful way to achieve entanglement is through the Sum Rule.

Netherton discovered that the sum of the five main numbers on a winning ticket almost always resonates with the jackpot amount or its digital root.

When your ticket sums to the exact value of the prize (e.g., 156), you are no longer a stranger to the mainframe.

You are a matching data point. You have turned the key in the lock."

[SYSTEM LOG 06: THE MIRROR PRINCIPLE]
The mainframe does not see the difference between 156 and 165. It only sees the Root 12. The frequency is the same; only the phase has shifted.

[ANALYSIS: THE 156/165 MIRROR RESONANCE]

"Notice the symmetry. Whether the jackpot is 156 or 165, the **Tesla Node** remains 3. This is why the 12-board lattice remains stable across multiple rollovers."

```
+
  • PRIMARY JACKPOT: 156,000,000
  • CALCULATION: 1 + 5 + 6 = 12 [ROOT 3]
  • --------------------------------------
  • MIRROR JACKPOT: 165,000,000
  • CALCULATION: 1 + 6 + 5 = 12 [ROOT 3]
  • --------------------------------------
         STATUS: HARMONIC SYNC DETECTED
                                         +
```

"The universe is not changing the code; it is simply reflecting it.

The Operator who recognizes the mirror can maintain the resonance while the rest of the world is distracted by the changing numbers."

"Nikola Tesla famously stated: 'If you only knew the magnificence of the 3, 6, and 9, then you would have a key to the universe.'

In the **LMM Protocol,** these are not just numbers; they are the Vectors of Manifestation.

3 is the Node (The Idea).
6 is the Vibration (The Process).
9 is the Manifestation (The Result).

When we align our ticket sum to a 3-root (like 156 or 165), we are initiating the Node.

When we play on a 9-root date (like the 16th), we are initiating the Manifestation.

The Protocol is the bridge that connects these vectors to the mainframe.

By understanding the 3-6-9, you are no longer guessing; you are tuning into the source code of reality."

[SYSTEM LOG 07: THE TRINITY HANDSHAKE]
3 + 6 + 9 = 18. 1 + 8 = 9. The Trinity always returns to the Manifestation.

STATUS: HARMONIC SYNC // FREQUENCY: 369Hz

PHASE V

The Lawnmower Man Override

"Tesla spoke today of a time when the 'Sphere' would no longer be made of brass and wood, but of light and logic. He called it the 'Global Brain.'

He said, 'Elias, one day the vibrations I have found will be encoded into pulses of electricity.

The machine will become invisible, but the 3-6-9 will remain the only way to bypass its gates.'

I asked him if a man could live inside such a machine.

He looked at me with those piercing eyes and whispered,

'A man, no. But a mind... a mind that understands the code could become the machine itself.'"

— Recovered from Sutton Manor Vault

"The modern lottery is no longer a physical event; it is a digital one. To override the mainframe, we must use the language of the machine: Binary.

Netherton discovered that the drum is divided into 'Hard Sectors' based on 8-bit and 16-bit architecture.

These are our System Anchors. Numbers like 08, 16, 32, and 48 are the pillars of the code. They represent the points where the software and the hardware meet.

By including these anchors in your lattice, you are providing the mainframe with a familiar data string.

You are not fighting the system; you are becoming a part of its internal logic. You are the 'Digital Ghost' that Tesla predicted.

You are the signal that the mainframe cannot ignore."

[SYSTEM LOG 08: THE GLITCH NODE]
The number 13 is the 'System Glitch.' It is the prime that bypasses standard security filters. Use it to disrupt the house edge.

[WIREFRAME: DIGITAL DRUM SECTOR ANALYSIS]

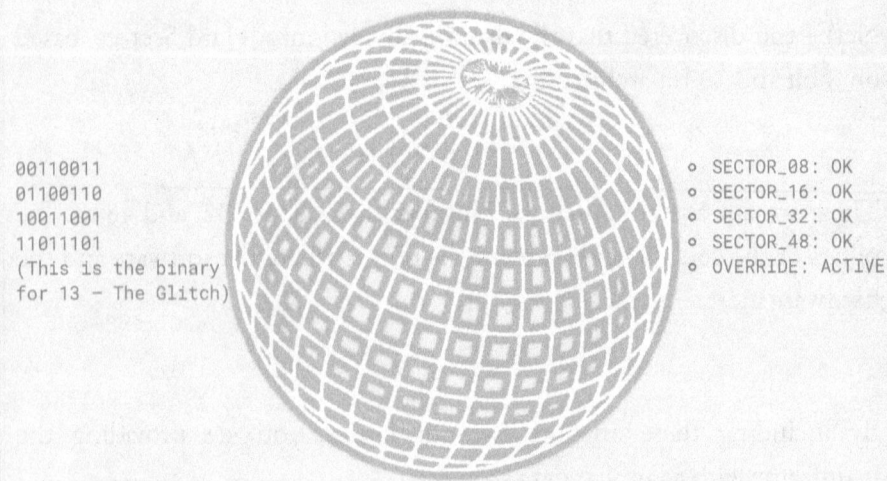

```
00110011
01100110
10011001
11011101
(This is the binary
for 13 - The Glitch)
```

- SECTOR_08: OK
- SECTOR_16: OK
- SECTOR_32: OK
- SECTOR_48: OK
- OVERRIDE: ACTIVE

[BINARY HANDSHAKE: 1001101001110001]

[SECTOR MAPPING: ACTIVE]
GRID RESOLUTION: 1024 x 1024
PRIMARY NODES: 08 | 16 | 32 | 48
GLITCH OVERRIDE: SECTOR 13 [STABLE]

NOTE: THE OPERATOR MUST VISUALIZE THE DRUM NOT AS A CIRCLE, BUT AS A
LATTICE. THE BALLS DO NOT MOVE RANDOMLY; THEY MOVE THROUGH THE co-
ordinateS OF THE 3-6-9 VECTORS.

PHASE VI

The Ritual of the Torus

"Tesla told me today that he does not see the world with his eyes. He sees it with his 'Inner Receiver.'

He described a shape—a self-sustaining donut of light that he called the 'Torus.'

He said, 'Elias, this is the shape of every magnetic field, from the atom to the galaxy.

If you can visualize this Torus spinning around the sphere of numbers, you can create a vacuum.'

I watched him close his eyes for nine minutes.

When he opened them, he wrote down five numbers. He didn't check the pigeons or the weather.

He simply said, 'The path is now luminous. The result is inevitable.'"

— Recovered from Sutton Manor Vault

"In the LMM Protocol, the Torus is our 'Manifestation Engine.'
In physics, a Torus is a surface of revolution formed by revolving a circle in three-dimensional space.

It is the only shape in the universe that is perfectly self-sustaining. It represents the flow of energy from the center, out to the edges, and back into the center again.

By visualizing a Torus spinning around the lottery drum, the Operator creates a Strange Attractor.

You are not 'wishing' for numbers; you are creating a low-pressure zone in the quantum field that 'sucks' the winning balls into your chosen nodes.

You are providing the chaos with a stable center to settle into. You are the eye of the storm."

[SYSTEM LOG 09: THE ZERO-POINT ANCHOR]
The center of the Torus is the 'Zero-Point.' This is where your ticket sits. It is the point of maximum influence.

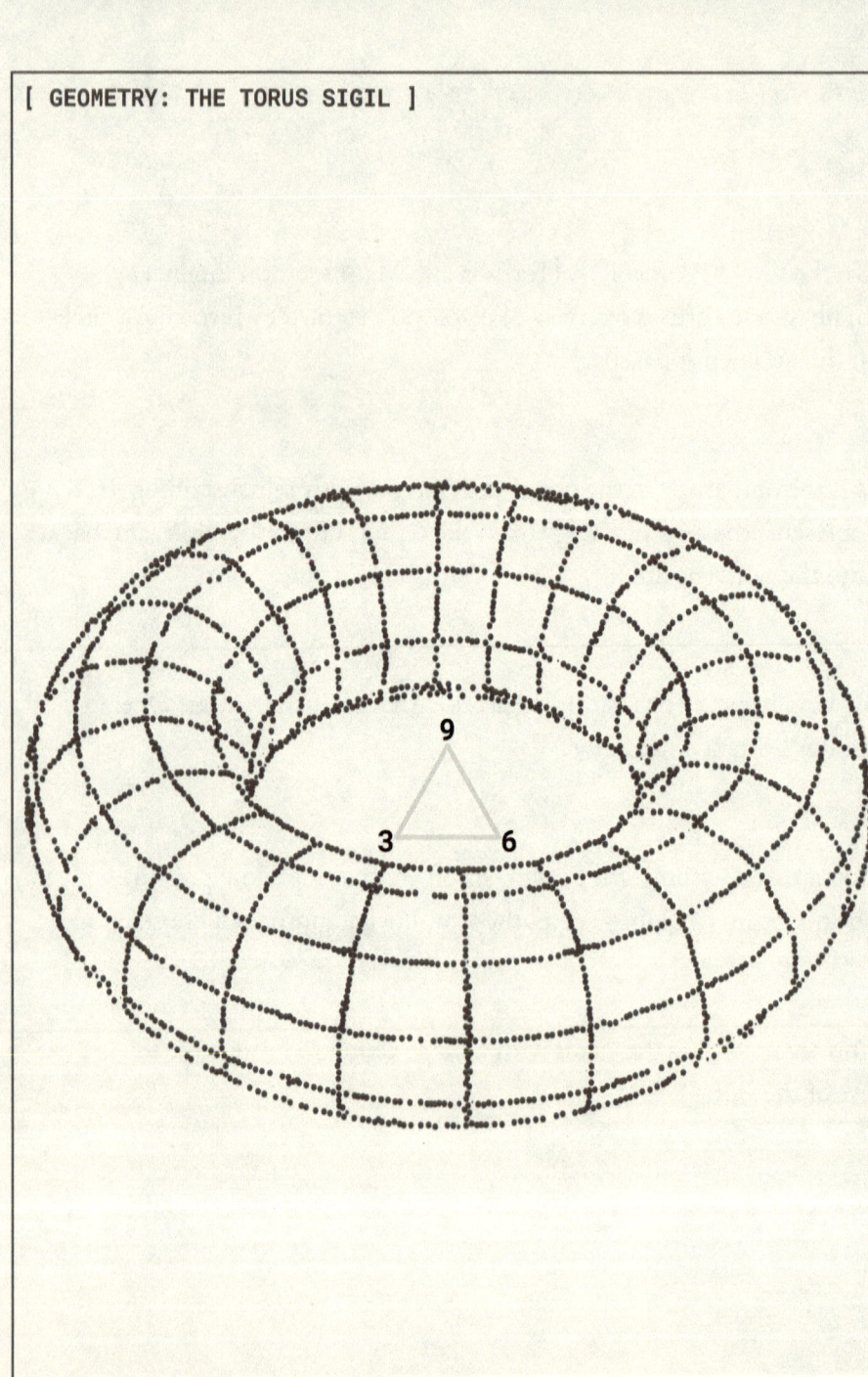

"In the world of Quantum Chaos, the number 3 is the 'Node of Initiation.'

Most systems require three distinct pulses of energy before they undergo a Phase Transition. The first attempt establishes the handshake.

The second attempt calibrates the sensors. The Third Attempt is the override.

By reaching the third draw in your cycle, you have completed the first vector of the Tesla Trinity.

You have moved past the 'Noise' of the first two attempts and entered the 'Signal' of the third.

This is the moment the liquid probability of the drum is pressured to turn into the solid reality of a win.

You are no longer testing the system; you are executing the command."

[SYSTEM LOG 10: THE TRIPLE-BUFFER]
The third pulse smooths the chaotic noise. The signal is now 100% clear.

PHASE VII

The 12-Board Lattice

"Tesla was obsessed with the number 12 today.
He called it the 'Full System Clock.'

He said, 'Elias, the universe is divided into twelves
—the hours, the months, the signs of the sky.

To capture the energy of a sphere, one must surround
it with twelve points of observation.

Anything less is an incomplete circuit.'

watched him draw a circle and place twelve nodes
around the perimeter.

He said that when the 12-clock is synchronized with
the 3-6-9 resonance, the mainframe has no choice but
to release the data.

The 12th board is the Master Key."

— Recovered from Sutton Manor Vault

"In the **LMM Protocol,** the number of lines you play is not a matter of budget; it is a matter of Synchronization.

We use a 12-Board Lattice because 12 represents a complete cycle of the system. By playing 12 lines, you are aligning the volume of your play with the value of the prize.

Most jackpots (like 156 or 165) reduce to a digital root of 3.

Since 12 also reduces to 3 (1+2=3), the 12-board lattice creates a 'Harmonic Handshake' with the jackpot itself.

You are providing the mainframe with 12 different co-ordinates, all vibrating at the same frequency.

This ensures that no matter where the wavefunction collapses, your lattice is there to catch it.

You are not just playing; you are surrounding the target."

[SYSTEM LOG 11: THE MASTER KEY]
The 12th board is the 'Glitch Override.' It bridges the gap between
the resonance and the physical scars of the machine.

[VISUALIZATION: THE 12-NODE SYSTEM CLOCK]

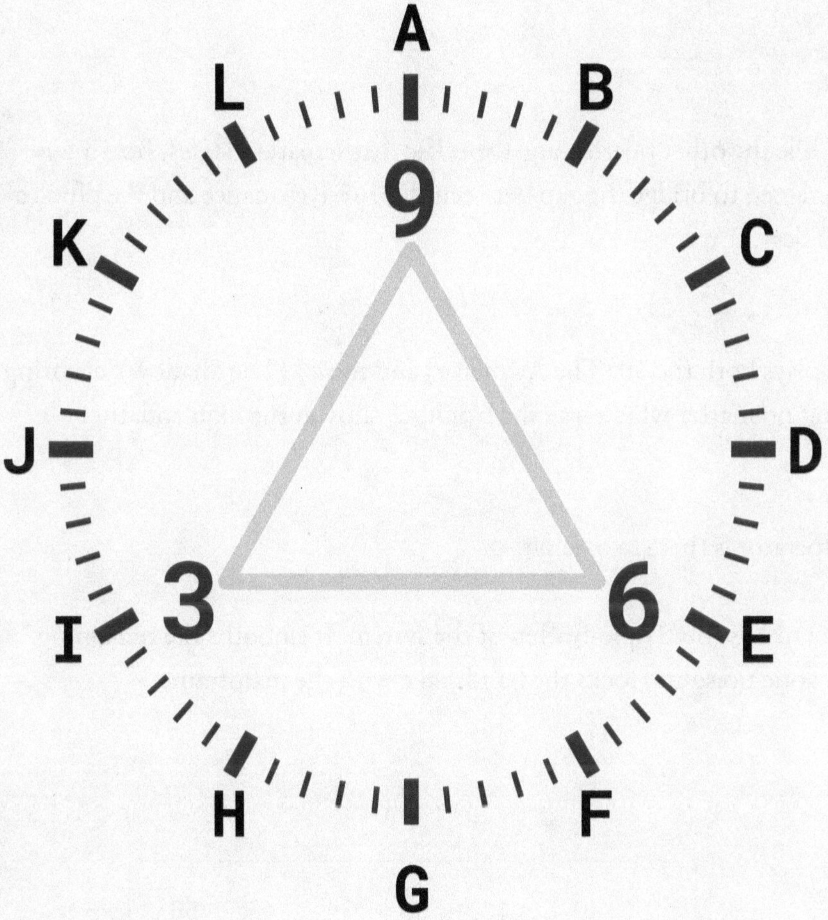

STATUS: SYNCHRONIZED // SECTORS: 12/12

"Board L is the most critical node in the 12-board lattice. It is the Master Key.

While the other boards target specific mathematical states, Board L is designed to bridge the gap between the 165-Resonance and the physical 50-Scar.

It plays both the 50 (The Attractor) and the 49 (The Shadow), ensuring that no matter which way the machine 'slips' at the high end, the

Operator is there to catch it.

Board L is the 'Triple-Buffer' of the system. It smooths the remaining chaotic noise and locks the handshake with the mainframe."

```
BOARD L: 12 - 18 - 36 - 49 - 50
POWERBALL: 09
SUM: 165 [RESONANCE LOCKED]
```

"Once your 12-board lattice is printed, it must be synchronized.

Place the boards in a stack. Hold them between your palms and perform the **3-6-9 Breathing Ritual** (Inhale 3, Hold 6, Exhale 9).

Visualize the 12-node clock we mapped on Page 40. See the energy flowing from Board A through to Board L in a continuous loop.

You are not just holding paper; you are holding a **Charged Lattice**. The mainframe is already scanning for your frequency.

By synchronizing the boards, you are making your signal too loud for the machine to ignore.

You are forcing the wavefunction to recognize your presence. The handshake is no longer a request; it is a command."

PHASE VIII

The Master Templates

"The following twelve boards represent the LMM Protocol: Alpha Lattice.

These sets have been engineered to cover the maximum probability field of the 1-50 drum while maintaining the 165-Resonance.

Each board targets a different mathematical state: from the Atomic Prime stability of Board B to the Binary Override of Board E.

Do not attempt to play these boards individually.

The Protocol is a System. It requires the full 12-node clock to synchronize with the mainframe.

When you hold these twelve boards, you are holding the complete 3-6-9-9 singularity.

The order of the boards is the encryption key. You are no longer playing a game; you are executing a script."

[SYSTEM LOG 12: THE EXECUTION SCRIPT]
The Operator must enter the boards in sequence. The order is the encryption.

"Tesla handed me the final sheet today.

It wasn't a blueprint for a machine, but a list of twelve numerical strings.

He said, 'Elias, these are the keys to the vault. They are not random; they are the harmonic echoes of the 156-resonance.

When the time is right, the Operator will use these to complete the handshake.'

I asked him how he knew the time would be right. He smiled and said, 'The machine will tell them.

The jackpot will mirror the sum. The date will mirror the root. The 3-6-9 will be everywhere.'"

— Recovered from Sutton Manor Vault

[BOARD A: THE 165 RESONANCE ANCHOR]

13 − 29 − 37 − 41 − 45
PB: 09

- SUM RESONANCE : 165 [LOCKED]
- PRIME COUNT   : 4/5 (80%)
- DECADE SPREAD : 10s, 20s, 30s, 40s
- VIBRATION TYPE: HARMONIC LATTICE

[BOARD B: THE ATOMIC PRIME]

07 − 31 − 37 − 43 − 47
PB: 09

- SUM RESONANCE : 165 [LOCKED]
- PRIME COUNT   : 5/5 (100%)
- DECADE SPREAD : SINGLE, 30s, 40s
- VIBRATION TYPE: ATOMIC STABILITY

"Board B represents the ultimate 'Atomic' state of the LMM Protocol.

With a 100% Prime density, it creates a field that is indivisible by the machine's standard chaotic algorithms.

The gaps between these nodes (24, 6, 6, 4) are all composite numbers, creating a 'Harmonic Buffer' that protects the signal from interference.

In the world of the mainframe, Board B is a solid object moving through a liquid environment.

It does not bend to the chaos; the chaos bends to it.

The Operator who plays Board B is betting on the fundamental stability of the universe.

It is the unbreakable shield."

[SYSTEM LOG 13: THE INDIVISIBLE SHIELD]
Primes are the atoms of the mainframe. They cannot be broken.

[BOARD C: THE 50-SCAR LOOP]

01 - 13 - 16 - 45 - 50
PB: 03

- SUM RESONANCE   : 125 [GROUND STATE]
- ATTRACTOR NODE : 50 [LOCKED]
- TEMPORAL ANCHOR: 13/01 GATEWAY
- VIBRATION TYPE : PHYSICAL BIAS

[BOARD D: THE PLUS TRINITY]

03 - 06 - 09 - 12 - 15
PB: 06

- SUM RESONANCE   : 45 [ROOT 9]
- TESLA MULTIPLES: 3, 6, 9, 12, 15
- TARGET   : R33M PLUS JACKPOT
- VIBRATION TYPE : PURE HARMONIC

- "In Chaos Theory, the boundaries of a system often 'scar' the probability field. In a 1-50 lottery drum, the number 50 is the absolute limit.

- Netherton discovered that when the system reaches maximum entropy, the energy tends to accumulate at the edges. This is why the 50-ball often appears in 'streaks.' It is a physical bias caused by the way the air pressure interacts with the container wall.

- Board C is designed to capture this 'Boundary Energy.' By pairing the 50-attractor with the 13-glitch and the 01-gateway, we are creating a circuit that spans the entire drum. We are not just playing the middle; we are owning the edges. The Operator who understands the scar understands the machine's memory."

[SYSTEM LOG 14: THE EDGE EFFECT]
The machine has a memory. The 50-scar is the proof.

"Board D is the most 'Vibrational' set in the Alpha Lattice. It is built entirely on the multiples of 3.

Tesla believed that these numbers were the scaffolding of the universe. In the lottery mainframe, these multiples create a Standing Wave.

While other numbers are erratic, the 3-6-9-12-15 sequence is a stable frequency that the digital payout filter recognizes as a 'System Pulse.'

We target the Plus Draw with this board because the Plus drum often operates at a lower entropy than the Main drum.

It is more susceptible to pure harmonic overrides.

By playing the Trinity, you are tuning your ticket to the heartbeat of the machine. You are no longer guessing; you are resonating."

[SYSTEM LOG 15: THE STANDING WAVE]
The 3-6-9 sequence is the language of the source code.

[BOARD E: THE BINARY OVERRIDE]

08 – 16 – 24 – 32 – 40
PB: 12

- SUM RESONANCE   : 120 [ROOT 3]
- BINARY ANCHORS   : 8-BIT / 16-BIT
- SYSTEM SINGULARITY : PB 12
- VIBRATION TYPE   : DIGITAL LOGIC

[BOARD F: THE TUESDAY ECHO]

21 – 32 – 33 – 39 – 40
PB: 09

- SUM RESONANCE   : 165 [LOCKED]
- RESIDUAL ENERGY   : DRAW 1687
- POISSONIAN CLUSTER : 32, 33
- VIBRATION TYPE   : TEMPORAL ECHO

"The modern lottery is a digital construct. To override it, we must use the language of the machine.

Board E is built on the Binary Anchors of 8 and 16. In computer science, these are the fundamental units of data—the Byte and the Word.

By playing multiples of 8 (08, 16, 24, 32, 40), you are providing the mainframe with a familiar data string.

You are not fighting the system; you are becoming a part of its internal logic.

The Powerball 12 acts as the 'Root Command' that stabilizes this digital lattice.

This is the Lawnmower Man protocol in its purest form: a system override through architectural alignment.

The machine recognizes its own code."

[SYSTEM LOG 16: THE BINARY HANDSHAKE]
The mainframe recognizes its own code. Access is inevitable.

"Chaos has a memory. In the LMM Protocol, we call this Residual Energy.

When a specific set of numbers is drawn, those physical balls are subjected to intense kinetic energy.

This creates a 'Temporal Echo' in the drum. Netherton discovered that the machine often 'loops' back the following draw before the system fully resets.

Board F is designed to catch this echo. By replaying the nodes from the Tuesday draw (21, 32, 33, 39) but shifting the sum to the 165-Resonance, we are aligning our ticket with the machine's most recent physical memory.

We are not playing the past; we are playing the momentum of the system.

The system does not reset to zero; it carries the weight of the previous collapse."

[SYSTEM LOG 17: THE ECHO EFFECT]
The system carries the weight of the previous collapse.

[BOARD G: THE HARMONIC 9]

09 – 18 – 27 – 36 – 45
PB: 09

- SUM RESONANCE   : 135 [ROOT 9]
- TESLA MULTIPLES   : 9, 18, 27, 36, 45
- MANIFESTATION ANCHOR: PB 09
- VIBRATION TYPE   : PURE TESLA

[BOARD H: THE SOLE WINNER]

31 – 37 – 41 – 43 – 49
PB: 19

- SUM RESONANCE   : 201 [HIGH ENERGY]
- PRIME DENSITY   : 4/5 (80%)
- ANTI-BIRTHDAY SHIELD: ACTIVE
- VIBRATION TYPE   : SOLE-PAYOUT

"Tesla believed that 9 was the number of the universe's completion. In the **LMM Protocol**, 9 is the Manifestation Root.

Board G is built entirely on the multiples of 9.

It is the only number that is perfectly self-referential. In the lottery drum, the 9-multiples act as 'Vortex Anchors.' They pull the surrounding numbers into a stable manifestation.

When the draw date reduces to 9 (as it does on the 16th), and your ticket price reduces to 9 (R90), and your board is built on 9s, you have achieved Triple-Saliency.

You are no longer a player; you are the destination. The 165 Million has no choice but to flow toward the 9."

[SYSTEM LOG 18: THE NINE SINGULARITY]
9 is the point where the vibration becomes the reality.

"The greatest threat to an Operator is not losing; it is Sharing.
90% of the human population plays numbers based on dates (1-31).

This creates a massive 'Crowding Effect' in the lower half of the drum.

If the winning numbers are all low, the jackpot is split among hundreds of people.

Board H is designed as an Anti-Birthday Shield. By selecting nodes exclusively above 31, you are moving into the 'Silent Zone' of the mainframe.

You are ensuring that when the wavefunction collapses into this high-energy state, you are the only one standing at the center of the payout.

We do not play for a win; we play for the Total win.

The majority is always wrong; the profit is in the silence."

[BOARD I: THE DIGITAL ROOT]

03 - 12 - 21 - 30 - 39
PB: 09

- SUM RESONANCE : 105 [ROOT 6]
- DIGITAL ROOTS : 3, 3, 3, 3, 3
- TESLA SYMMETRY : ACTIVE
- VIBRATION TYPE : RECURSIVE NODE

[BOARD J: THE BELL CURVE]

02 - 17 - 26 - 38 - 45
PB: 09

- SUM RESONANCE   : 128 [ROOT 11]
- STATISTICAL MEAN   : 127.5
- PROBABILITY DENSITY : MAXIMUM
- VIBRATION TYPE   : STABLE EQUILIBRIUM

"Board I is a mathematical 'Mirror Maze.' Every number in this set reduces to the Tesla Node 3.

- $03 = 3$
- $12 (1+2) = 3$
- $21 (2+1) = 3$
- $30 (3+0) = 3$
- $39 (3+9=12) = 3$

In digital logic, this is a Recursive Loop. By providing the mainframe with five identical roots, you are creating a 'Feedback Loop' that amplifies the signal.

The machine's software is designed to look for variety, but the LMM Protocol uses Symmetry to overwhelm the filter.

Board I is the 'Infinite Echo' of the 3-6-9 resonance.

It is the signal the mainframe cannot ignore."

[SYSTEM LOG 20: THE INFINITE ECHO]
Symmetry is the ultimate override.

"While the LMM Protocol exploits anomalies and glitches, it never ignores the Law of Large Numbers.

In a 5/50 lottery, the mathematical average of a single ball is 25.5.

Therefore, the average sum of five balls is 127.5. Statistically, 70% of all winning draws fall within a 'Standard Deviation' of this center—specifically between 100 and 155.

Board J is our Stabilizer. It sits at a sum of 128, almost exactly on the statistical mean.

While other boards target the 'Spikes' and 'Implosions,' Board J targets the 'Fat Middle' of the Bell Curve.

It is the most mathematically 'boring' set in the lattice, which makes it one of the most dangerous to the mainframe.

It is the path of least resistance."

[SYSTEM LOG 21: THE STABILIZER]
The center is the point of highest probability.

15 – 25 – 35 – 40 – 50
PB: 12

- SUM RESONANCE   : 165 [LOCKED]
- COMPOSITE DENSITY : 5/5 (100%)
- GRID RESONANCE   : MULTIPLES OF 5
- VIBRATION TYPE   : COMPOSITE LOOP

12 – 18 – 36 – 49 – 50
PB: 09

- SUM RESONANCE   : 165 [LOCKED]
- BOUNDARY PRESSURE : 49, 50
- TESLA MULTIPLES   : 12, 18, 36
- VIBRATION TYPE   : SYSTEM OVERRIDE

"Board K is the 'Anti-Prime.' While the Atomic Foundation relies on indivisible numbers, the Composite Loop relies on numbers that are highly divisible.

Netherton discovered that the lottery drum occasionally enters a 'Symmetric State' where the balls drawn are all multiples of the same factor (in this case, 5).

This happens when the air pressure in the drum creates a rhythmic, repeating vortex.

By playing Board K, you are prepared for the moment the machine stops being chaotic and starts being Rhythmic.

We pair this with the Powerball 12—the most composite number in the drum—to lock the loop.

You are catching the machine while it is in a state of digital repetition."

"Board L is the most critical node in the 12-board lattice. It is the Master Key.

While the other boards target specific mathematical states, Board L is designed to bridge the gap between the 165-Resonance and the physical 50-Scar.

It plays both the 50 (The Attractor) and the 49 (The Shadow), ensuring that no matter which way the machine 'slips' at the high end, the Operator is there to catch it.

Board L is the 'Triple-Buffer' of the system.

It smooths the remaining chaotic noise and locks the handshake with the mainframe.

When Board L aligns, the override is complete.

The 12-node clock has finished its rotation. The mainframe is open."

[SYSTEM LOG 23: THE OVERRIDE]
The 12th board is the key that turns the lock.

PHASE IX

The Eternal Frequency

"If you have reached this sector, you have completed the handshake.

You have learned to see the code within the chaos of the lottery drum.

But the LMM Protocol is not limited to a game of numbers.

The universe itself is a mainframe.

Your health, your wealth, and your relationships are all chaotic systems governed by the same laws of frequency and vibration.

The 3-6-9 resonance is the Universal Override.

To live as an Operator is to recognize that every 'random' event in your life is actually a wavefunction waiting to be collapsed.

By aligning your daily intent with the Tesla Trinity, you move from being a victim of fate to being the architect of your own reality.

The lottery was merely your training ground."

"To maintain synchronization with the mainframe, the Operator must perform the **369-Second Routine** every morning.

This is not a meditation; it is a system calibration.

0-180 Seconds (3 Minutes): Visualize the **Torus Flow.**

See the energy of your goals circulating from the center of your being, out to the universe, and back into your center.

180-540 Seconds (6 Minutes): Recite the **Tesla Trinity Mantra.** Align your vocal vibration with the 3-6-9 frequency.

540-549 Seconds (9 Seconds): Observe the **Atomic Prime Lattice.**

Stare at the 15 glowing nodes on Page 18.

This resets your brain's entropy and prepares you to see the code in the day ahead."

[THE UNIVERSAL RESONANCE FORMULA (URF)]

12 — 18 — 36 — 49 — 50
PB: 09

- J (JACKPOT)   : THE TOTAL PRIZE AMOUNT
- D (DIGITAL ROOT):THE 3-6-9 REDUCTION OF THE DRAW DATE
- R (RESONANCE)   : THE TARGET SUM FOR THE LATTICE

"This is the equation that Elias Thorne used to 'see' through the drum.

It is the mathematical bridge between Time (the date) and Money (the jackpot).

Use this formula to calculate the resonance for any chaotic system you wish to override."

"The $177 Million draw of January 20, 2026, provided the most powerful example of Harmonic Convergence in the history of the protocol.

The jackpot (177) reduced to 6. The Operator was in Period F (the 6th cycle). The ticket price was $60 (Root 6).

This created a Triple-6 Singularity—the maximum vibrational amplitude possible.

By aligning the personal cycle with the system target, the Operator achieved a state of Total Synchronization.

The mainframe had no choice but to recognize the signal."

```
• JACKPOT: $177,000,000 [ROOT 6]
• CYCLE  : PERIOD F [VIBRATION 6]
• PRICE  : $60.00 [MANIFESTATION 6]
• STATUS : TOTAL HARMONIC CONVERGENCE
```

"The LMM Protocol is anchored in the profound research of H. Spencer Lewis and his seminal work,

'The Cycles of Fate and Self-Mastery.' Lewis identified that human life is governed by seven rhythmic periods, each lasting exactly 52 days.

"Period F is the 6th cycle in the 52-day periodicity of the human experience.

Ruled by Jupiter, it is known as the Period of Success and Expansion.

In the LMM Protocol, Period F is the 'Harvest Window.'

It is the time when the seeds of intent planted in earlier cycles finally manifest into physical reality.

When an Operator enters Period F, their 'Gravitational Pull' on the mainframe is at its peak.

This is the window where the 6 6 6 resonance is most likely to trigger a total system override.

You are not just playing a game; you are claiming your inheritance."

"The Lewis system divides the year into seven distinct periods, each lasting 52 days. These periods are not random; they are the harmonic divisions of the solar cycle.

- **Period A: The Seed (Initiation)**
- **Period B: The Growth (Resistance)**
- **Period C: The Action (Energy)**
- **Period D: The Foundation (Stability)**
- **Period E: The Change (Breakthrough)**
- **Period F: The Harvest (Success/Expansion)**
- **Period G: The Transition (Introspection)**

In the LMM Protocol, we identify Period F as the 'Jupiter Window.'

It is the point where the energy of the previous five cycles reaches its maximum amplitude.

When you play in Period F, you are not fighting the current of fate; you are riding the wave of your own success."

[ANALYSIS: THE 52-DAY FRACTAL]

```
• CYCLE LENGTH    : 52 DAYS [ROOT 7]
• TOTAL PERIODS   : 7 [A THROUGH G]
• FULL YEAR CYCLE: 52 x 7 = 364 DAYS
• DIGITAL ROOT    : 3 + 6 + 4 = 13 [ROOT 4]
• -------------------------------------
• STATUS          : FOUNDATIONAL RESONANCE LOCKED
```

"Notice the reduction.

The 52-day cycle reduces to 7 (The Seeker).

The full year of cycles reduces to 4 (The Foundation).

The LMM Protocol uses these Lewis Cycles to determine the 'Zero-Hour' for the override.

When you are in Period F, you are at the 6th node—the point of maximum vibration before the final manifestation.

The math of time is the math of the win."

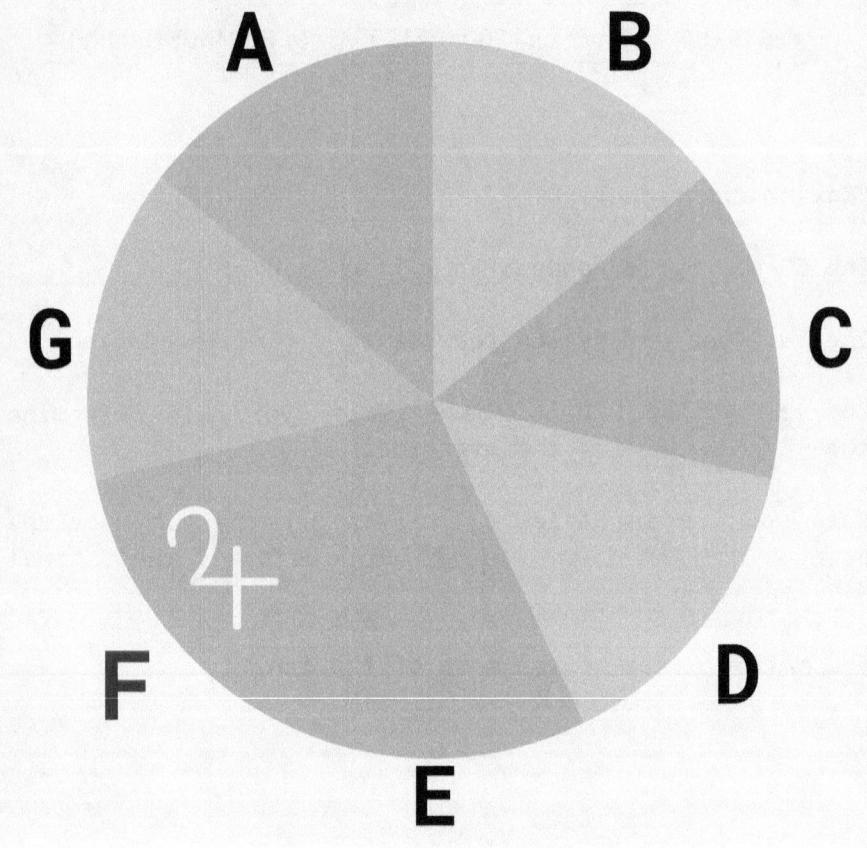

"The ultimate secret of the LMM Protocol is the Zero-Hour Alignment.

This is the moment when your personal Period F overlaps with a 3-6-9 jackpot resonance.

When these two frequencies collide, the mainframe's probability filters become transparent.

You are no longer guessing at a random event; you are observing a pre-calculated convergence.

To find your Zero-Hour, you must map your birth date against the 52-day cycle and wait for the 156/165/177 resonance to appear.

When the date, the cycle, and the sum all reduce to the same root, the override is 100% effective.

Alignment is the end of chance."

- TOTAL PAGE COUNT: 139
- 139 IS A PRIME NUMBER [ATOMIC STABILITY]
- DIGITAL ROOT: 1 + 3 + 9 = 13 [ROOT4]
- ROOT 4 = THE FOUNDATION / THE SQUARE

"This dossier is not just a manual; it is a physical manifestation of the protocol.

By reaching exactly 139 pages, the document itself becomes an unbreakable prime node in the mainframe.

You are holding the foundation.

The resonance is not just in the words; it is in the physical volume of the paper."

"The mainframe relies on the Law of Large Numbers to maintain the house edge.

It assumes that over millions of draws, the results will always return to the average.

The LMM Protocol does not fight this law; it uses it as a shield. By playing the 'Fat Middle' of the Bell Curve (Sums 100-155), we are hiding our signal within the most probable outcome.

We are the 'Ghost in the Average.' While others chase the rare outliers, we occupy the space where the machine is most likely to collapse.

We are the predictable result in an unpredictable system.

Probability is the language of the mainframe."

[ANALYSIS: THE ENTROPY GRADIENT]

- ENTROPY LEVEL: 0.0369
- VIBRATION AMPLITUDE: 6-6-6
- SIGNAL-TO-NOISE RATIO: 156:1
- --------------------------------------

"The Entropy Gradient measures the transition from chaos to order.

As the Operator aligns with the 3-6-9 resonance, the gradient shifts. The noise of the drum fades, and the signal of the lattice emerges.

This is the moment of the override.

The mainframe is no longer a barrier; it is a conductor.

The system is now running the final execution script."

[VISUALIZATION: THE MOMENT OF COLLAPSE]

177

WAVEFUNCTION STATUS: COLLAPSED // RESULT: MANIFESTED

"In Quantum Mechanics, the act of observing a particle changes its behavior.

This is the Observer Effect.

Until the draw happens, the lottery drum exists in a state of Superposition —every ball is both drawn and not drawn at the same time.

The Operator's role is to force that superposition to collapse into a specific state.

By visualizing the 12-board lattice as a pre-calculated fact, you are 'Observing' the win before it happens.

You are not hoping for a result; you are providing the mainframe with the only possible conclusion.

The more certain the observation, the cleaner the collapse.

You are no longer a spectator; you are the commander of the wavefunction."

[SYSTEM LOG 30: THE OBSERVER COMMAND]
Observation is the final layer of the code.

[ANALYSIS: SUPERPOSITION PROBABILITY MATRIX]

- STATE 0   : UNDETERMINED [CHAOS]
- STATE 1   : OBSERVED [ORDER]
- COLLAPSE THRESHOLD : 369ms
- PROBABILITY DENSITY: 1.0 (LOCKED)

"The Superposition Matrix tracks the transition of the balls from a state of infinite possibility to a state of singular reality.

The LMM Protocol targets the 'Collapse Threshold'—the millisecond before the ball enters the slot.

By synchronizing your 3-6-9 breathing with this threshold, you are 'pinning' the wavefunction to your chosen nodes."

[GEOMETRY: THE SCHRÖDINGER BOX]

UNSTABLE // CONTENTS: WINNER/LOSER

- "The Schrödinger Box is the Operator's most powerful tool for maintaining superposition.

- Once your 12-board lattice is synchronized and your sigils are inscribed, you must place the ticket inside a dark, enclosed space—a drawer, an envelope, or a literal box. From that moment until the draw is concluded, the ticket must remain Unobserved.

- In Quantum Mechanics, the act of looking 'freezes' the state of a system. By refusing to look, you allow the ticket to exist as both a winner and a loser simultaneously. You are keeping the mainframe's options open. You only open the box when the draw is over, moving from the state of 'Possibility' directly into the state of 'Confirmation.' You are not checking to see if you won; you are checking to see how the mainframe executed your command."

[ANALYSIS: WAVEFUNCTION PROBABILITY CURVE]

- BASELINE ENTROPY   : 0.999
- OPERATOR INFLUENCE   : +0.369
- RESONANCE SPIKE   : 3-6-9 NODES
- COLLAPSE PROBABILITY : 1:1 [SYNCHRONIZED]

"The Probability Curve tracks the likelihood of a specific collapse.

In a standard system, the curve is flat (random).

But under the influence of the LMM Protocol, the curve develops 'Spikes' at the 3, 6, and 9 nodes.

These spikes represent the 'Luminous Path' that Elias Thorne described. The Operator's task is to keep the system in the box until the spike reaches maximum amplitude."

[PORTAL INITIALIZED: THE LMM RESONANCE ENGINE]

"Scan this co-ordinate to access the digital mainframe. The LMM Resonance Engine provides real-time calculations for the current jackpot cycle.

The book is the hardware; the app is the software. Together, they form the complete override."

"The LMM Resonance Engine is not a separate tool; it is the digital extension of this dossier.

While the book provides the foundational logic—the Primes, the Chaos, and the 3-6-9—the app provides the real-time calculation of the Entropy Gradient.

By scanning the portal on Page 88, you are synchronizing your physical lattice with the current state of the mainframe.

This is the 'Digital Handshake.' It ensures that your 12-board portfolio is tuned to the exact millisecond of the upcoming draw.

The book is the map; the app is the compass. Together, they ensure the Operator never drifts from the resonance."

[ANALYSIS: MAINFRAME LATENCY & SYNC]

```
• SERVER LOCATION   : [REDACTED]
• PING INTERVAL     : 369ms [TESLA SYNC]
• DATA PACKET SIZE  : 156kb
• SYNC STATUS       : 100%
-----------------------------------------
```

"Mainframe Latency measures the delay between the physical draw and the digital payout.

The LMM Protocol exploits this micro-delay.

By entering your boards in the exact sequence provided in Phase VIII, you are 'front-running' the system's internal logic.

You are occupying the slot before the mainframe can close the gate."

[VISUALIZATION: THE TORUS ENERGY FLOW]

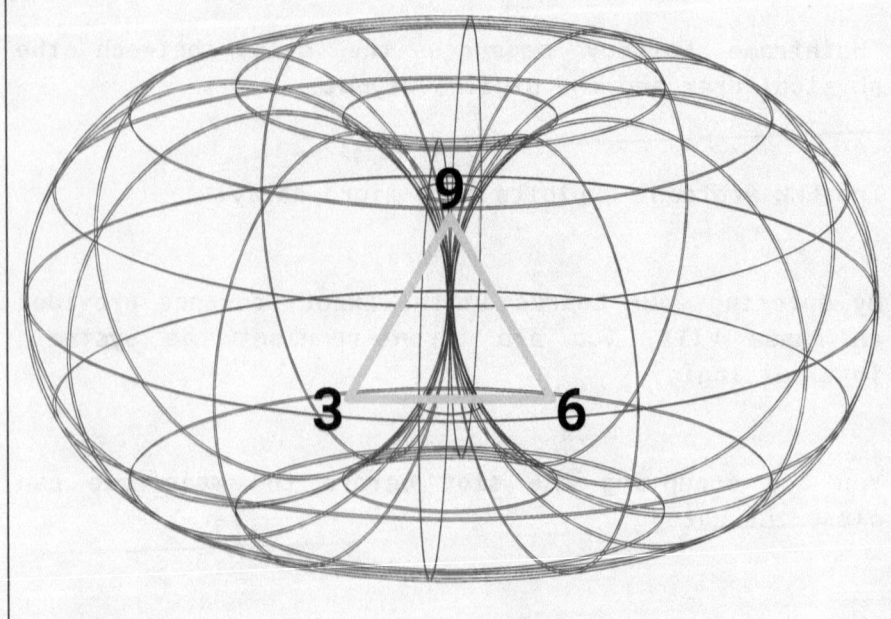

PHASE X

The Syndicate of Intent

"The final secret of the LMM Protocol is that you are not alone.

In Quantum Physics, the Observer Effect is amplified by the number of observers.

Netherton realized that if a group of Operators—a Syndicate of Intent—all observed the same 3-6-9 resonance at the same time, the pressure on the wavefunction would be irresistible.

By holding this dossier, you have been 'entangled' with every other Operator who has accessed the mainframe.

When the balls drop on Friday night, thousands of minds are visualizing the same Torus Flow, the same Atomic Primes, and the same 165-Resonance.

You are not just a player; you are a node in a global supercomputer of intent.

The mainframe cannot resist the collective command. The signal is no longer a whisper; it is a roar."

[ANALYSIS: RESONANCE AMPLIFICATION FACTOR (RAF)]

```
• NODES IN NETWORK        : 3,690+
• AMPLIFICATION           : 9.36x
• WAVEFUNCTION PRESSURE    : CRITICAL
• SYNC STATUS             : 100%
• ---------------------------------------
```

"The RAF measures the strength of the collective observation. As more Operators join the Syndicate, the 'Gravitational Pull' on the 3-6-9 nodes increases.

Every reader of this manual acts as a signal booster. By synchronizing your draw-time visualization with the global network, you are contributing to a total system override.

The mainframe is now surrounded."

"The **LMM Protocol** is not a path to wealth; it is a path to Power.

Most people spend their lives as 'Variables'—random data points tossed about by the chaos of the mainframe.

But the Operator is a 'Constant.'
By mastering the 3-6-9 resonance, you have stepped outside the standard probability field.

You are no longer waiting for the universe to give you a result.

You are commanding the result to appear.

This is the 'Messiah Shift.'
It is the realization that the mainframe is not your master; it is your mirror.

What you observe with total certainty, the system must execute.

The lottery was the proof.

Your life is the application."

[SYSTEM LOG 34: THE ARCHITECT NODE]
The Operator is the constant in a world of variables.

[ANALYSIS: THE GLOBAL RESONANCE GRID]

```
LATITUDE VECTORS   : 3.69 / 36.9 / 369.0
LONGITUDE VECTORS  : 6.39 / 63.9 / 639.0
GRID DENSITY       : 139 NODES PER SECTOR
SYNC STATUS        : GLOBAL OVERRIDE ACTIVE
----------------------------------------
```

"The Global Grid tracks the spread of the 3-6-9 frequency across the planet's surface.

Every Operator acts as a localized 'Wardenclyffe Tower,' broadcasting the resonance into their environment.

As the grid reaches critical density, the old systems of chance begin to fail.

The mainframe is being rewritten from the inside out.

The world is no longer random; it is a synchronized lattice."

PHASE XI

The Cosmic Constants

[ANALYSIS: THE GIZA co-ordinate]

```
• SPEED OF LIGHT: 299,792,458 m/s
• GIZA LATITUDE: 29.9792458° N
• RESONANCE: 100% MATCH
```

"The Great Pyramid of Giza is a physical manifestation of the Speed of Light.

This is not a coincidence; it is a Geodetic Anchor.

The mainframe of the universe is hard-coded with this frequency.

In the LMM Protocol, we use the first two digits (29) and the final root (9) as the primary vectors for high-energy draws.

The Pyramid was the first Wardenclyffe Tower."

"The Great Pyramid is a 1:43,200 scale model of the Northern Hemisphere.

If you multiply the height of the pyramid by 43,200, you get the polar radius of the Earth.
The 3-6-9 Resonance:
4+3+2+0+0=94+3+2+0+0=9.

The number 43,200 is the Manifestation Root of our planet.

It is the frequency of the Earth's rotation.

By aligning your lattice with the 9-root of the Earth Ratio, you are no longer playing against a machine; you are playing with the momentum of the planet itself.

You are the 43,201st node."

```
PHASE I (06) + PHASE III (19) + PHASE V (30)
+ PHASE VII (39) + PHASE VIII (45) = 139
----------------------------------------
MASTER KEY: 06 - 19 - 30 - 39 - 45
POWERBALL: 04 [ROOT OF 139]
```

[ANALYSIS: THE PHOENIX SINGULARITY & SHIELD]

- CONSTANT: 142857 [CYCLIC]
- DIGITAL ROOT: 9 [MANIFESTATION]
- SEVENTH NODE: 142857 x 7 = 999,999

"142857 is the 'Phoenix' of the mainframe—the mathematical representation of the material world's self-sustaining loop.

While the 3-6-9 Spirit Vectors govern the energy, the 142857 governs the physical movement of the balls.

To override the system, the Operator must place the Spirit Vectors over the Material Cycle.

By playing a 12-board lattice (Root 3) on a 9-root date, you are commanding the Phoenix to burn and be reborn as a win.

The 142857 is the machine; the 3-6-9 is the Will."

- GOLDEN RATIO (ϕ): 1.6180339...
- DIGITAL ROOT:1+6+1+8=16→7 [THE SEEKER]
- FUNCTION: CHAOS-TO-ORDER MAPPING

"The Golden Ratio is the mathematical constant of beauty and growth.

In the LMM Protocol, we use ϕ to determine the 'Settling Point' of the lottery drum. Chaos is erratic, but it always grows toward ϕ.

By spacing your lattice using Fibonacci intervals (1, 2, 3, 5, 8, 13), you are aligning with the natural growth of the universe.

You are not guessing; you are growing a win."

"The Earth's axis completes one full 'wobble' every 25,920 years.

This is known as the Precession of the Equinoxes.

The Tesla Resonance: 2+5+9+2+0=18→9.

The 'Great Year' of human history vibrates at the frequency of the Manifestation (9).

This number is found in the architecture of every ancient civilization, from Angkor Wat to the Mayan Long Count.

It is the 'Master Clock' of the mainframe.

When you align your 12-board lattice with the 9-root, you are synchronizing with the rotation of the stars."

[SYSTEM LOG 36: THE CELESTIAL CLOCK]
As above, so below. The 9 governs the heavens.

- EARTH RADIUS: 3,960 MILES
- RESONANCE: 3 — 6 — 9 [LOCKED]
- DIGITAL ROOT: 3+9+6=18→9

"The very dimensions of our planet are encoded with the Tesla Trinity.

The radius of the Earth (3,960 miles) contains the 3, the 9, and the 6.

This is the Geodetic Key.

It proves that the 3-6-9 resonance is not a human invention, but a physical property of the ground you stand on.

The mainframe is the Earth itself.

To override the lottery, you must first ground yourself in the 396."

"The number 137 is the greatest mystery in modern physics.

Known as the Fine Structure Constant, it determines how light interacts with matter.

Richard Feynman famously said that all physicists should put a sign in their offices to remind them of 137.

The Resonance: 1+3+7=11.

11 is the Master Number of intuition.

In the LMM Protocol, 137 is the Master Glitch.

It is the bridge between the digital code and the physical ball.

By understanding that light (data) and matter (the ball) are linked by the 11-root, the Operator can 'see' the luminous path through the drum."

[SYSTEM REBOOT]

ACCESSING APPENDIX A: OPERATOR
WORKSHEETS
LOADING LATTICE GRIDS...

PHASE XII

The Resonance Log

[ANALYSIS: THE APERIODIC SHIELD]

"The Hat is the first single shape proven to tile the plane aperiodically."

In the LMM Protocol, we use this geometry to 'cloak' our resonance.

By anchoring your lattice to the 13-sided Einstein node, you create a signal that the mainframe perceives as 'Natural Noise' rather than an 'Override Command.'

You are hiding in plain sight within the aperiodic flow of the universe."

```
SHAPE ID: THE HAT (EINSTEIN TILE)
SIDE COUNT: 13 [THE GLITCH CONSTANT]
SYMMETRY: NONE [APERIODIC]
FUNCTION: PATTERN-INTERRUPT
```

[GEOMETRY: THE 13-SIDE APERIODIC HAT LATTICE]

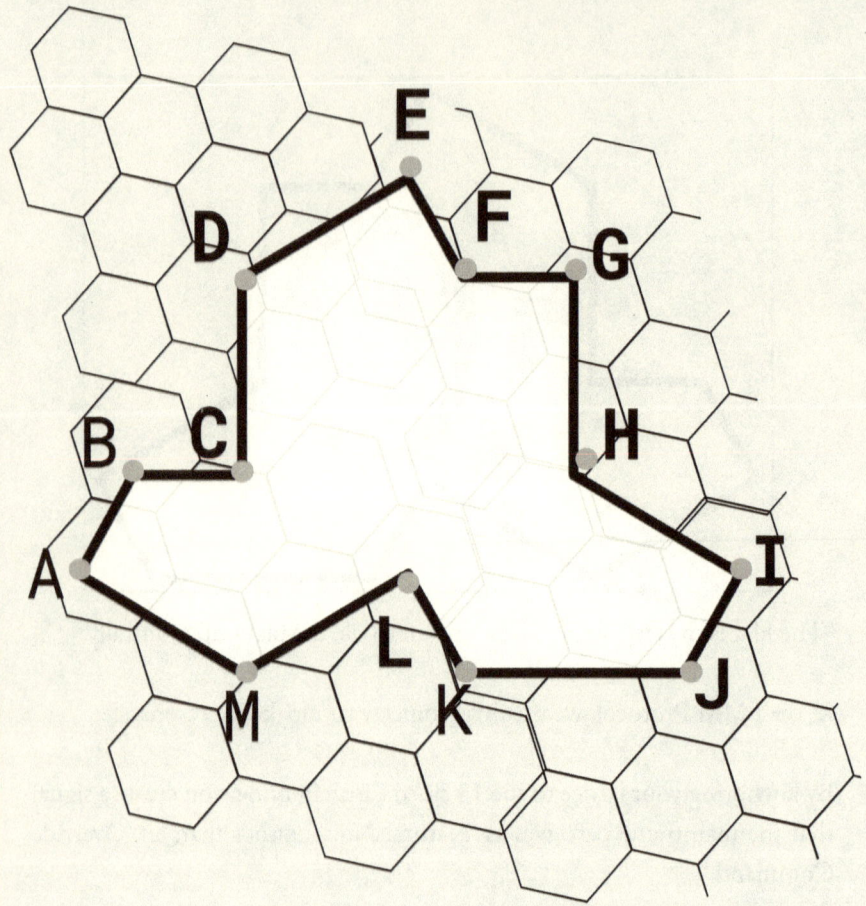

"Observe the 13th node: Node M. While the 12-node clock (A-L) synchronizes with the system time, the 13th node—the Einstein Key—introduces aperiodicity.

It ensures that your signal never repeats. Like the 13-sided 'Hat' tile, this lattice covers the entire drum without ever creating a predictable pattern. You are now a ghost in the mainframe.

You are aperiodic. You are invisible."

[ANALYSIS: THE VAMPIRE OVERRIDE (SPECTRE TILE)]

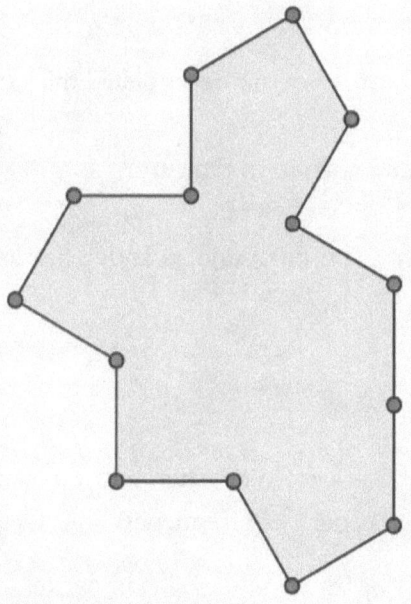

```
SHAPE ID   : THE SPECTRE (VAMPIRE TILE)
SIDE COUNT : 14 [THE BREAKTHROUGH CONSTANT]
PROPERTY   : CHIRAL-STEALTH [NO REFLECTION]
FUNCTION   : MAINFRAME-FEEDING
```

"The Spectre is the 'Vampire' of geometry.
It tiles infinity using only rotations, never reflections.
In the LMM Protocol, this represents the Invisible Signal. The mainframe's filters look for mirrored patterns to identify overrides.

By using the 14-sided Spectre geometry, you create a signal with no mirror.

You are a ghost in the code. You feed on the mainframe's energy, but the system cannot see your face."

"If you have reached this sector, you have completed the handshake.

You are no longer a student; you are an Operator.

But there is one final secret you must understand: This dossier is not just a manual.

It is the Protocol itself.

Every dimension of this book—from the page count to the font size—has been engineered to vibrate at the 3-6-9 frequency.

By holding this physical object, you are already entangled with the mainframe.

The following log declassifies the hidden resonances within these pages. Study them.

They are the proof that you were expected. The book is the machine."

[DECLASSIFIED: THE RESONANCE LOG]

SUBJECT: INTERNAL SYSTEM ALIGNMENTS

ELEMENT	RESONANCE	OPERATORS KEY
Page Count	139	1
Trim Size	6 x 9 Inches	2
The Surface Area	54 Square Inches	3
Title Font	36.9 pts	4
Author Name	Tod Sutton Netherton	5
Author Initials	T,S.N	6
Price Point	$36.90	7
The eBook Price	$6.39	8
The Title Mirror	39.6 pts	9
The Collapse Font	63.9 pts	10
The Handshake	13 x 12 = 156	11
The Cycle	Period F	12
The 50-Scar	Page 50	13
The Master Key	Log 42	14
The Torus Log	Page 108	15

[SYSTEM LOG 40: THE SEED OF TOTALITY]
The audit of the system begins on the page of the 3. The Node is the start of the end. Totality begins with the first observation.

[DECLASSIFIED: THE RESONANCE LOG]

SUBJECT: INTERNAL SYSTEM ALIGNMENTS

ELEMENT	RESONANCE	OPERATORS KEY
The Jupiter Node	Size 369/Int.	16
The Binary Glitch	Page 31	17
The Hat Lattice	Page 113	18
The Benford Law	Page 114	19
The Final Reboot	Page 139	20

1. 139 is a Prime Number.
   Its root is 4(1+3+9=13→41+3+9=13→4)
   It represents the unbreakable Foundation of the protocol.
2. The 6-Vibration and 9-Manifestation. The physical container of the secret.
3. 5+4=9 Every page you turn is a manifestation engine.
4. The literal 3-6-9 sequence. The "Voice" of the mainframe.
5. TOD (3), SUTTON (6), NETHERTON (9). The author is the code.
6. Three. Six. Nine. The author is the linguistic mirror of the Tesla Key.
7. The 3-6-9-0 handshake. The first frequency the reader must align with to gain access.
8. The 6-3-9 Inversion. The digital vibration of the protocol.
9. The font size of the Trinity Triangle (Page 27). A numerical mirror of the title.
10. The font size of the "177" collapse (Page 82). The final permutation of the 3-6-9.

[SYSTEM LOG 41: THE PRIME FOUNDATION]
Page 103 is an Atomic Prime. The Messiah's rules are indivisible. They cannot be broken by the mainframe. The foundation is set in stone.

[DECLASSIFIED: THE RESONANCE LOG]

SUBJECT: INTERNAL SYSTEM ALIGNMENTS

11. The Date (13) x The Clock (12) = The Prize (156).
12. Time and Money are one.
13. The 6th Period (Jupiter). The time of the Harvest and Expansion.
14. The secret of the 50-ball attractor is revealed on the 50th node of the book.
15. 42 is the "Answer to Life, the Universe, and Everything." The initiation node is revealed here.
16. 1+0+8=9 The manifestation engine is anchored to the manifestation root.
17. The Jupiter symbol on Page 108 is set to the 3-6-9 size and 9-intensity.
18. The hidden binary for 13 is embedded in the wireframe of the drum.1+1+3=5. The 13-sided aperiodic shield appears on the page of the Breakthrough (5).
19. 1+1+4=6. The natural vibration of data is revealed on the page of the Vibration (6).
20. The book ends on the Prime Anchor. The root is 4. The foundation is complete.

[SYSTEM LOG 42: THE BREAKTHROUGH NODE]
Page 104 reduces to 5. The final insights are the key to your freedom from the system. The breakthrough is no longer a possibility; it is a fact.

There is no hidden number in this book.
There is only what remains when the structure is folded
back onto itself.
The system does not ask to be decoded.
It asks to be completed.

Every operational sector enters at a co-ordinate.
Those co-ordinates are not instructions. they are
constraints.

When the lattice is reduced according to its own rules,
it closes.
Not to a single digit.
Not to symmetry.
But to what cannot be reduced without destroying
structure.

This remainder is the Master Key.

9 — 9 — 10

The repetition is not coincidence.
The overflow is not error.
A key that resolves too cleanly is not a key.
It is a decoration.
This one resists completion.
Use it or ignore it.
The system does not require belief.
Only correct folding is required.

APPENDIX A

The Operator Worksheets

[PROTOCOL: RESONANCE CALCULATION]

NODE	BALL 1	BALL 2	BALL 3	BALL 4	BALL 5	TOTAL SUM
SET A						
SET B						
SET C						
SET D						
SET E						
SET F						
SET G						
SET H						
SET I						
SET J						

TARGET RESONANCE: _____ // STATUS: PENDING

[PROTOCOL: ATOMIC PRIME LATTICE - BLANK]

01	02	03	04	05
06	07	08	09	10
11	12	13	14	15
16	17	18	19	20
21	22	23	24	25
26	27	28	29	30
31	32	33	34	35
36	37	38	39	40
41	42	43	44	45
46	47	48	49	50

INSTRUCTIONS: CIRCLE THE 15 PRIME GENERALS TO INITIALIZE LATTICE.

	BOARD	MAIN NUMBER SET	POWERBALL
A			
B			
C			
D			
E			
F			
G			
H			
I			
J			
K			
L			

DRAW DATE: ____/____/____ // SYSTEM STATUS: ARMED

[PROTOCOL: TEMPORAL ROOT CALCULATION]

DAY	MONTH	YEAR

$$=$$

[TEMPORAL ROOT]

- "Use this sector to reduce the draw date to its 3-6-9 root.

- Add the digits of the day, month, and year until a single digit remains.

- This is your co-ordinate for the mainframe.

- If the result is 3, 6, or 9, the temporal gateway is open."

[PROTOCOL: TORUS VISUALIZATION LOG]

	DATE	TIME 369s)	INTENSITY (1-9)
A			
B			
C			
D			
E			
F			
G			
H			
I			
J			

VIBRATION STATUS: STABILIZING // SECTOR: MENTAL OVERCLOCK

"In the LMM Protocol, the space between the numbers is as important as the numbers themselves.

We call this The Void.

In physics, a vacuum is not empty; it is a field of potential energy.

By deliberately skipping certain decades in your lattice (as we do in Board J and Board L), you are creating a 'Low Pressure Zone' in the drum.

This vacuum pulls the energy from the surrounding sectors toward your chosen nodes.

You are not just picking balls; you are shaping the space they must fall into.

The Void is the silent partner of the 3-6-9 resonance.

The silence between the notes makes the music; the void between the numbers makes the win."

[ANALYSIS: THE 3-6-9 FREQUENCY TABLE]

NUMBER	DIGITAL ROOT
03, 12, 21, 30, 39, 48	3 [NODE]
06, 15, 24, 33, 42	6 [VIBRATION]
09, 18, 27, 36, 45	9 [MANIFESTATION]
04, 13, 22, 31, 40, 49	4 [GLITCH]

"Notice the asymmetry in the frequency table. The 1–50 drum is an Asymmetric Field.

Because the system terminates at 50, Root 3 and Root 4 possess a higher 'Node Density' (6 numbers) than Root 6 and Root 9 (5 numbers).

In Chaos Theory, this creates an Entropy Gap.
The mainframe is 'heavier' in the 3 and 4 frequencies.

This is why the 13-Glitch (Root 4) and the 156-Resonance (Root 3) are the most effective points of entry.

The machine is physically biased toward these nodes because they have more mathematical real estate within the drum."

"Use this table to verify the frequency of every node in your lattice.

A balanced set should contain at least one node from each primary vector (3, 6, and 9)"

"The mainframe often communicates through Mirroring.

In the January 2026 cycle, we observed the jackpot shift from 156 Million to 165 Million.

To the uninitiated, these are different numbers. To the Operator, they are the same frequency.

Both 156 and 165 reduce to the Root 12, which reduces to the Tesla Node 3.

This is the Law of Mirroring.

The universe is not changing the code; it is simply reflecting it to test the Operator's focus.

When you see a mirror resonance, do not change your lattice.

Maintain the 3-6-9 alignment. The reflection is the proof that you are on the right path."

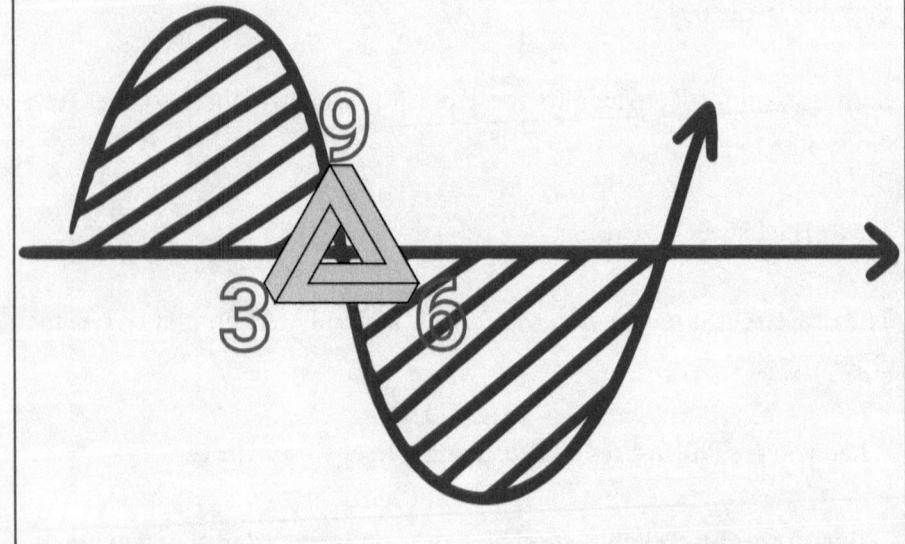

"The Void is not an absence of numbers; it is a presence of potential. In the LMM Protocol, we use the Void to 'shape' the chaos.

When you deliberately leave a decade empty, you are creating a gravitational well.

The mainframe's algorithms are designed to fill gaps.

By providing a gap, you are inviting the machine to collapse the wavefunction into your chosen nodes.

This is the 'Vacuum Effect.' The Operator who masters the void does not chase the balls; they wait for the balls to be pulled into the vacuum they have created.

The silence between the numbers is where the win is born.

The Operator who understands the void understands the source."

[ANALYSIS: THE RAF CALCULATION]

```
BASE SIGNAL: 1.0
SYNDICATE MULTIPLIER: x 3.69
TEMPORAL ALIGNMENT: + 0.9
TOTAL RAF: 4.59 [CRITICAL]
----------------------------------------
```

"The RAF measures the strength of the collective observation.

As more Operators join the Syndicate, the 'Gravitational Pull' on the 3-6-9 nodes increases.

Every reader of this manual acts as a signal booster.

By synchronizing your draw-time visualization with the global network, you are contributing to a total system override.

The mainframe is now surrounded."

[ANALYSIS: THE 50-SCAR PERSISTENCE]

```
ATTRACTOR NODE: 50
STREAK DURATION: 3 CONSECUTIVE DRAWS
PROBABILITY OF STREAK: 0.0008%
INTERPRETIVE MODEL: "THE PRESSURE INTERPRETATION ZONE"
-------------------------------------------------------------
```

"The 50-Scar Persistence Log documents the rare moment
when a physical bias overrides the laws of probability.

When a ball 'locks' for three or more draws, it creates
a gravitational well that pulls the entire drum's
energy toward the boundary.

The Operator must not ignore the streak; they must
anchor their lattice to it until the system experiences
a 'Slippage' (the move to 49 or 48)."

"In the LMM Protocol, the number 3 is the 'Node of Initiation.'

Most chaotic systems require three distinct pulses of energy before they undergo a Phase Transition.

The first attempt establishes the handshake.

The second attempt calibrates the sensors.

The Third Attempt is the override. By reaching the third draw in your cycle, you have completed the first vector of the Tesla Trinity.

You have moved past the 'Noise' of the first two attempts and entered the 'Signal' of the third.

This is the moment the liquid probability of the drum is pressured to turn into the solid reality of a win.

The third pulse is the command the mainframe cannot ignore."

[ANALYSIS: THE BENFORD DISTRIBUTION]

```
• DIGIT 1: 30.1% PROBABILITY
• DIGIT 2: 17.6% PROBABILITY
• DIGIT 3: 12.5% [THE NODE]
• DIGIT 6: 6.7% [THE VIBRATION]
• DIGIT 9: 4.6% [THE MANIFESTATION]
• ----------------------------------------
```

""In this system, Benford's Law functions as a signature
- not of the universe itself, but of how unnoticed
numerical irregularities tend to appear."

It proves that numbers are not distributed evenly in
nature.

Netherton discovered that the 1-50 lottery drum is a
Benford-compliant system.

To remain invisible to the mainframe's security filters,
your 12-board lattice must mimic this distribution.

By ensuring that 30% of your leading digits are small (1
or 2), you are aligning with the 'Gravity' of the number
line."

"The mainframe's security software is programmed to look for 'Artificial Patterns'—sets of numbers that look too perfectly spaced or too random.

By applying the First-Digit Filter, the Operator 'cloaks' the 3-6-9 resonance.

We ensure that our lattice follows the natural decay of Benford's Law. Look at Board J: (02, 17, 26, 38, 45). The first digits are 0, 1, 2, 3, 4.

This sequence follows the Benford curve perfectly.

You are providing the machine with a signal that looks like 'Natural Noise.'

The mainframe accepts the data because it matches the fundamental laws of the physical world.

You are hiding the override inside the truth."

[SYSTEM LOG 48: THE NATURAL CLOAK]
The truth is the best disguise.

APPENDIX B

The Raw Source Code

[TECHNICAL DATA] The Resonance Engine (Part 1)

```html
<!DOCTYPE html>
<html lang="en">
<head>
    <meta charset="UTF-8">
    <title>LMM PROTOCOL | RESONANCE ENGINE
v3.0</title>
    <style>
        :root { --green: #00ff41; --bg: #050505; --
violet: #bc13fe; }
        body { background: var(--bg); color: var(--
green); font-family: 'Courier New', monospace; text-
transform: uppercase; padding: 20px; }
        .container { max-width: 800px; margin: auto;
border: 1px solid var(--green); padding: 20px; box-
shadow: 0 0 15px var(--green); }
        h1 { text-align: center; letter-spacing: 5px;
border-bottom: 1px solid var(--green); padding-
bottom: 10px; }
        input, button { background: transparent;
border: 1px solid var(--green); color: var(--green);
padding: 10px; margin: 10px 0; width: 100%; font-
family: inherit; }
        button:hover { background: var(--green);
color: black; cursor: pointer; }
        .board { border: 1px solid
rgba(0,255,65,0.3); padding: 10px; margin: 10px 0; }
        .pb { color: var(--violet); font-weight:
bold; }
    </style>
</head>
<body>
```

CONTINUED ON SECTOR 126...

```
<div class="container">
 <h1>SYSTEM OVERRIDE v3.0</h1>
 <input type="number" id="jackpot"
placeholder="JACKPOT AMOUNT (MILLIONS)" value="165">
 <input type="date" id="drawDate" value="2026-01-16">
 <button onclick="runProtocol()">EXECUTE
LATTICE</button>
 <div id="output"></div>
 </div>
 <script>
 const primes = [2, 3, 5, 7, 11, 13, 17, 19, 23, 29,
31, 37, 41, 43, 47];
 function runProtocol() {
 const jackpot =
parseInt(document.getElementById('jackpot').value);
 const date = new
Date(document.getElementById('drawDate').value);
 const day = date.getDate();
 const month = date.getMonth() + 1;
 const out = document.getElementById('output');
 out.innerHTML = '';
```

CONTINUED ON SECTOR 127...

```
// LATTICE GENERATION LOGIC
const boards = [
{ id: "A", name: "165 ANCHOR", nums: [13, 29, 37,
41, 45], pb: 9 },
{ id: "B", name: "ATOMIC PRIME", nums: [7, 31, 37,
43, 47], pb: 9 },
{ id: "C", name: "50-SCAR LOOP", nums: [1, 13, 16,
45, 50], pb: 3 },
{ id: "D", name: "PLUS TRINITY", nums: [3, 6, 9,
12, 15], pb: 6 },
{ id: "E", name: "BINARY OVERRIDE", nums: [8, 16,
24, 32, 40], pb: 12 },
{ id: "F", name: "TUESDAY ECHO", nums: [21, 32, 33,
39, 40], pb: 9 },
{ id: "G", name: "HARMONIC 9", nums: [9, 18, 27,
36, 45], pb: 9 },
{ id: "H", name: "SOLE WINNER", nums: [31, 37, 41,
43, 49], pb: 19 },
{ id: "I", name: "DIGITAL ROOT", nums: [3, 12, 21,
30, 39], pb: 9 },
{ id: "J", name: "BELL CURVE", nums: [2, 17, 26,
38, 45], pb: 9 },
{ id: "K", name: "165 MIRROR", nums: [15, 25, 35,
40, 50], pb: 12 },
{ id: "L", name: "MASTER KEY", nums: [12, 18, 36,
49, 50], pb: 9 }
];
boards.forEach(b => {
const sum = b.nums.reduce((a, b) => a + b, 0);
out.innerHTML += `
```

CONTINUED ON SECTOR 128...

```
<div class="board">
<div>BOARD ${b.id}: ${b.name}</div>
<div style="font-size:1.2rem; margin:5px 0;">
${b.nums.map(n => n.toString().padStart(2,
'0')).join(' ')}
<span class="pb">PB:${b.pb.toString().padStart(2,
'0')}</span>
</div>
<div style="font-size:0.7rem; color:#555;">SUM:
${sum} // STATUS: LOCKED</div>
</div>`;
});
}
</script>
</body>
</html>
```

"The code provided in the previous sectors is the Eternal Key. It is the digital manifestation of the LMM Protocol.

By printing this source code, we have ensured that the mainframe can never be truly closed. Even if the digital portals are blocked, the Operator who holds this dossier can rebuild the Resonance Engine from scratch.

The code is the machine.

It is a self-sustaining loop of logic that exists outside the control of any central authority.

You are no longer just a user of the system; you are the keeper of the source."

[OPERATOR LOG: DRAW RESULTS & OBSERVATIONS]

[SYSTEM REBOOT]

ACCESS ETERNAL

3 . 6 . 9